~A BINGO BOOK~

# Elementary Science Bingo Book

## COMPLETE BINGO GAME IN A BOOK

Pu'u 'O'o, a Volcanic cone on Kilauea, Hawaii
Source: United States Geological Survey

**Written By Rebecca Stark**

**Educational Books 'n' Bingo**

ISBN 978-0-87386-463-3

**Educational Books 'n' Bingo**

Printed in the U.S.A.

# BINGO DIRECTIONS

**INCLUDED:**

List of Terms

Templates for Additional Terms and Clues

2 Clues per Term

30 Unique Bingo Cards

Markers

1. **Either cut apart the book or make copies of ALL the sheets. You might want to make an extra copy of the clue sheets to use for introduction and review. Keep the sheets in an envelope for easy reuse.**

2. Cut apart the call cards with terms and clues.

3. Pass out one bingo card per student. There are enough for a class of 30.

4. Pass out markers. You may cut apart the markers included in this book or use any other small items of your choice.

5. Decide whether or not you will require the entire card to be filled. Requiring the entire card to be filled provides a better review. However, if you have a short time to fill, you may prefer to have them do the just the border or some other format. Tell the class before you begin what is required.

6. There are 50 terms. Read the list before you begin. If there are any terms that have not been covered in class, you may want to read to the students the term and clues before you begin.

7. There is a blank space in the middle of each card. You can instruct the students to use it as a free space or you can write in answers to cover terms not included. Of course, in this case you would create your own clues. (Templates provided.)

8. Shuffle the cards and place them in a pile. Two or three clues are provided for each term. If you plan to play the game with the same group more than once, you might want to choose a different clue for each game. If not, you may choose to use more than one clue.

9. Be sure to keep the cards you have used for the present game in a separate pile. When a student calls, "Bingo," he or she will have to verify that the correct answers are on his or her card AND that the markers were placed in response to the proper questions. Pull out the cards that are on the student's card keeping them in the order they were used in the game. Read each clue as it was given and ask the student to identify the correct answer from his or her card.

10. If the student has the correct answers on the card AND has shown that they were marked in response to the *correct questions,* then that student is the winner and the game is over. If the student does not have the correct answers on the card OR he or she marked the answers in response to *the wrong questions,* then the game continues until there is a proper winner.

11. If you want to play again, reshuffle the cards and begin again.

## Have fun!

# TOPICS INCLUDED

air

amphibian

autumn

birds

blood

brain

clouds

day

earth

earthquake

electricity

energy

equator

fish

hair

hearing

heart

heat

insects

light

lungs

magnet

mammals

moon

mountain

nerves

ocean

planet

plant

precipitation

reptiles

rocks

seasons

sight

simple machines

skin

smell

sound

spring

star

summer

sun

taste

temperature

touch

volcano

water

wind

winter

year

# Additional Terms

Choose as many Science terms as you would like and write them in the squares. Repeat each as desired. Cut out the squares and randomly distribute them to the class. Instruct the students to place the square on the center space of their card.

|  |  |  |  |  |
|---|---|---|---|---|
|  |  |  |  |  |
|  |  |  |  |  |
|  |  |  |  |  |
|  |  |  |  |  |
|  |  |  |  |  |
|  |  |  |  |  |

Elementary Science Bingo

# Clues for Additional Terms

Write three clues for each of your Science terms.

_____

1.

2.

3.

_____

1.

2.

3.

_____

1.

2.

3.

_____

1.

2.

3.

_____

1.

2.

3.

_____

1.

2.

3.

| | |
|---|---|
| **AIR**<br><br>1. It is a mixture of gases we need for breathing.<br><br>2. An invisible mixture of gases, mostly oxygen and nitrogen. | **AMPHIBIAN**<br><br>1. These animals begin life with gills and then develop lungs.<br><br>2. A frog is one. It begins life as a tadpole. |
| **AUTUMN**<br><br>1. It is the season that follows summer.<br><br>2. In the United States this season starts in September. | **BIRDS**<br><br>1. These are the only animals with feathers.<br><br>2. Robins and parrots are two kinds. |
| **BLOOD**<br><br>1. It is the fluid pumped through the body by the heart.<br><br>2. This fluid circulates through the body and carries oxygen and nourishment. | **BRAIN**<br><br>1. This organ of the body is enclosed in the skull.<br><br>2. This organ of the body controls almost all we do. |
| **CLOUDS**<br><br>1. They are made up of droplets of water and ice crystals.<br><br>2. Cumulus, stratus and cirrus are three types of ___. | **DAY**<br><br>1. There are 24 hours in one ___. That is the length of time it takes Earth to revolve around the sun.<br><br>2. Earth's rotation causes ___ and night. |
| **EARTH**<br><br>1. It is the name of our planet. Like the other rocky ones, it is divided into 4 layers: the inner core, the outer core, the mantle and the crust.<br><br>2. Like the other planets, it revolves around the sun. | **EARTHQUAKE**<br><br>1. It is the shaking and vibration of the earth.<br><br>2. It is the violent shaking of the earth's crust that sometimes causes destruction. |

Elementary Science Bingo

| | |
|---|---|
| **ELECTRICITY**<br><br>1. Energy created by movement of charged particles, such as electrons and protons.<br><br>2. Lightning is a natural form of this. | **ENERGY**<br><br>1. The capacity to do work.<br><br>2. Heat, light and sound are three forms of this. |
| **EQUATOR**<br><br>1. The imaginary line around the center of the Earth.<br><br>2. It is always in regions near this. | **FISH**<br><br>1. This kind of animal has gills all of its life.<br><br>2. This kind of animal can only survive in water. |
| **HAIR**<br><br>1. The threadlike growth that covers our heads.<br><br>2. Most mammals are covered in it. | **HEARING**<br><br>1. One of our senses. The others are smell, taste, touch and vision.<br><br>2. The ears are the organs for this sense. |
| **HEART**<br><br>1. This organ pumps blood through the body,<br><br>2. This organ is really a large muscle. It is part of our circulatory system. | **HEAT**<br><br>1. This form of energy is transferred by a difference in temperature.<br><br>2. We get ___ energy and light energy from the sun. |
| **INSECTS**<br><br>1. They have six legs. All are divided into three main parts: a head, a thorax and an abdomen.<br><br>2. There are more than a million species. Ants, bees and grasshoppers are three examples. | **LIGHT**<br><br>1. It makes sight possible.<br><br>2. We get ___ energy and heat energy from the sun. |

Elementary Science Bingo

| | |
|---|---|
| **LUNGS**<br><br>1. We could not breathe without these organs of the body.<br><br>2. These organs are part of the respiratory system. | **MAGNET**<br><br>1. This kind of object attracts iron or steel.<br><br>2. Opposite poles of a ___ will attract. Like poles will repel. |
| **MAMMALS**<br><br>1. These warm-blooded animals give birth to live young. They include humans.<br><br>2. Dogs, cats, rabbits, horses and elephants are all examples. | **MOON**<br><br>1. It is another word for a natural satellite that revolves around a planet.<br><br>2. Earth has only one ___. Some planets have several. |
| **MOUNTAIN**<br><br>1. It is a land mass that rises up above its surroundings.<br><br>2. Mount Everest on the Nepal-Tibet border is the tallest ___. | **NERVES**<br><br>1. These bundles of fibers form a network for conducting information throughout the body.<br><br>2. Along with the brain and spinal cord, they are part of the nervous system. |
| **OCEAN**<br><br>1. It covers more than 70% of Earth.<br><br>2. A continuous body of water that is divided into areas known as the Pacific, the Atlantic, the Indian and the Southern. | **PLANETS**<br><br>1. They orbit around the sun. Earth is one.<br><br>2. There are eight: Mercury, Venus, Earth, Mars, Jupiter, Saturn, Neptune and Uranus. Pluto used to be classified as one. |
| **PLANT**<br><br>1. It is a living organism that is not an animal.<br><br>2. The basic parts of a typical one are roots, stems, leaves, flowers, fruits, and seeds. | **PRECIPITATION**<br><br>1. When clouds become too heavy, the particles fall to Earth as ___.<br><br>2. Some types of ___ are rain, snow, and hail. |

Elementary Science Bingo

| | |
|---|---|
| **REPTILES**<br><br>1. These cold-blooded animals are covered in scales. The females lay eggs.<br>2. Crocodiles, snakes, lizards and turtles are all ___. | **ROCKS**<br><br>1. They are made up of minerals.<br>2. Three kinds are sedimentary, igneous and imetamorphic. Examples of each are limestone, quartz and marble. |
| **SEASONS**<br><br>1. There are four: winter, spring, summer and autumn.<br>2. The ___ are caused because Earth is tilted on its axis as it revolves around the sun. | **SIGHT**<br><br>1. It is one of our senses. The others are sight, taste, touch and vision.<br>2. The eyes are our organs for this sense. |
| **SIMPLE MACHINES**<br><br>1. They make work easier for us.<br>2. There are six kinds: pulley, lever, wedge, wheel and axle, screw, and inclined plane. | **SKIN**<br><br>1. This organ acts as a protective covering for our bodies.<br>2. This organ is important for our sense of touch. |
| **SMELL**<br><br>1. Our nose is the organ we use for this sense.<br>2. Complete this analogy:<br><br>    eyes : sight :: nose : ___<br><br>(Read *eyes* is to *sight* as *nose* is to ___.) | **SOUND**<br><br>1. This kind of energy is caused by vibrations.<br>2. It produces the sensation of hearing. |
| **SPRING**<br><br>1. Plants begin to grow during this season in the Temperate Zones. It is not too hot and not too cold.<br>2. In the United States this season begins in March and ends in June. | **STAR**<br><br>1. Our sun is one.<br>2. A ___ gives off its own light, but a planet does not. |

Elementary Science Bingo

© Barbara M. Peller

| | |
|---|---|
| **SUMMER**<br><br>1. It is the warmest season of the year.<br><br>2. In the United States and other parts of the Northern Hemisphere this season begins in June and ends in September. | **SUN**<br><br>1. It is a star; the planets in our solar system orbit around it.<br><br>2. Solar energy is energy from the ___. |
| **TASTE**<br><br>1. This sense lets us distinguish among salty, sour, bitter, and sweet.<br><br>2. The tongue and throat are the parts of the body we use for this sense. | **TEMPERATURE**<br><br>1. It is the degree of hotness or coldness.<br><br>2. We use a thermometer to measure this. |
| **TOUCH**<br><br>1. It is one of our senses. The others are sight, taste, smell and hearing.<br><br>2. The skin is the organ for this sense. | **VOLCANO**<br><br>1. It is a mountain where liquid rock erupts through the surface of the planet.<br><br>2. When one erupts, lava, ash, cinders, dust, and hot gas can pour out of the vent at the top. |
| **WATER**<br><br>1. This clear, colorless, tasteless liquid is necessary for life.<br><br>2. In its solid state it is called ice. In its gas state it is called steam. | **WIND**<br><br>1. Moving air.<br><br>2. Windmills are powered by ___ energy. |
| **WINTER**<br><br>1. It is the coldest season of the year.<br><br>2. In the United States and other parts of the Northern Hemisphere this season begins in December and ends in March. | **YEAR**<br><br>1. It is the length of time it takes Earth to revolve around the sun.<br><br>2. There are 365 days in a normal one. |

Elementary Science Bingo

© Barbara M. Peller

# Elementary Science Bingo

| Mammals | Insects | Temperature | Winter | Volcano |
|---|---|---|---|---|
| Day | Air | Wind | Reptiles | Magnet |
| Star | Smell | | Mountain | Rocks |
| Year | Simple Machines | Heart | Summer | Moon |
| Nerves | Energy | Earth | Hearing | Heat |

Elementary Science Bingo: Card No. 1

# Elementary Science Bingo

| Year | Taste | Planets | Seasons | Nerves |
|------|-------|---------|---------|--------|
| Moon | Reptiles | Amphibian | Simple Machines | Spring |
| Skin | Energy | | Earthquake | Heart |
| Hair | Sun | Smell | Ocean | Magnet |
| Heat | Wind | Earth | Day | Hearing |

# Elementary Science Bingo

| Year | Heart | Reptiles | Summer | Star |
|------|-------|----------|--------|------|
| Energy | Air | Brain | Insects | Lungs |
| Simple Machines | Wind | | Spring | Autumn |
| Smell | Skin | Nerves | Hair | Planets |
| Hearing | Day | Earth | Ocean | Temperature |

# Elementary Elementary Science Bingo

| | | | | |
|---|---|---|---|---|
| Smell | Spring | Nerves | Day | Temperature |
| Plant | Amphibian | Insects | Seasons | Star |
| Mountain | Hair | | Volcano | Winter |
| Heart | Sound | Wind | Earth | Brain |
| Light | Heat | Precipitation | Hearing | Rocks |

# Elementary Science Bingo

| Heat | Volcano | Simple Machines | Amphibian | Day |
|---|---|---|---|---|
| Plant | Heart | Brain | Earthquake | Air |
| Taste | Rocks | | Electricity | Equator |
| Magnet | Spring | Mammals | Ocean | Light |
| Reptiles | Earth | Sound | Smell | Mountain |

# Elementary Science Bingo

| Autumn | Spring | Planets | Taste | Rocks |
|---|---|---|---|---|
| Summer | Simple Machines | Light | Insects | Star |
| Seasons | Brain | | Amphibian | Earthquake |
| Earth | Nerves | Ocean | Precipitation | Temperature |
| Moon | Heart | Mammals | Mountain | Sound |

# Elementary Science Bingo

| Mammals | Spring | Equator | Electricity | Reptiles |
|---------|--------|---------|-------------|----------|
| Moon | Temperature | Energy | Air | Plant |
| Planets | Winter | | Earthquake | Birds |
| Smell | Hair | Star | Year | Skin |
| Earth | Day | Ocean | Precipitation | Autumn |

# Elementary Science Bingo

| Mountain | Spring | Clouds | Summer | Birds |
|----------|--------|--------|--------|-------|
| Plant | Taste | Seasons | Rocks | Amphibian |
| Star | Sight | | Temperature | Volcano |
| Hearing | Smell | Year | Light | Hair |
| Wind | Earth | Precipitation | Simple Machines | Moon |

# Elementary Science Bingo

| | | | | |
|---|---|---|---|---|
| Earthquake | Reptiles | Energy | Star | Rocks |
| Light | Taste | Mountain | Simple Machines | Temperature |
| Lungs | Mammals | | Air | Clouds |
| Sight | Heat | Nerves | Electricity | Equator |
| Hair | Ocean | Brain | Year | Volcano |

Elementary Science Bingo: Card No. 9

# Elementary Science Bingo

| Year | Summer | Amphibian | Seasons | Sound |
|------|--------|-----------|---------|-------|
| Rocks | Birds | Insects | Air | Temperature |
| Sight | Spring | | Winter | Skin |
| Nerves | Magnet | Light | Ocean | Lungs |
| Blood | Moon | Planets | Heat | Mountain |

© Barbara M. Peller

# Elementary Science Bingo

| Autumn | Spring | Simple Machines | Light | Moon |
|--------|--------|-----------------|-------|------|
| Clouds | Lungs | Electricity | Earthquake | Insects |
| Plant | Taste | | Planets | Energy |
| Blood | Star | Ocean | Day | Year |
| Brain | Earth | Mammals | Precipitation | Reptiles |

# Elementary Science Bingo

| | | | | |
|---|---|---|---|---|
| Reptiles | Volcano | Lungs | Summer | Earthquake |
| Energy | Wind | Taste | Precipitation | Air |
| Mammals | Equator | | Rocks | Seasons |
| Earth | Hair | Temperature | Year | Plant |
| Spring | Clouds | Sight | Brain | Birds |

# Elementary Science Bingo

| | | | | |
|---|---|---|---|---|
| Blood | Volcano | Autumn | Lungs | Rocks |
| Taste | Clouds | Spring | Earthquake | Skin |
| Summer | Amphibian | | Energy | Equator |
| Mountain | Ocean | Birds | Sight | Year |
| Earth | Magnet | Precipitation | Mammals | Electricity |

# Elementary Science Bingo

| Day | Taste | Simple Machines | Earthquake | Blood |
|-----|-------|-----------------|------------|-------|
| Birds | Mammals | Lungs | Air | Spring |
| Light | Winter | | Planets | Brain |
| Magnet | Ocean | Sight | Amphibian | Autumn |
| Earth | Seasons | Skin | Moon | Mountain |

Elementary Science Bingo: Card No. 14

# Elementary Science Bingo

| Electricity | Earthquake | Simple Machines | Reptiles | Summer |
|---|---|---|---|---|
| Autumn | Planets | Insects | Taste | Light |
| Rocks | Mammals | | Star | Temperature |
| Earth | Lungs | Clouds | Ocean | Blood |
| Moon | Hair | Precipitation | Sound | Energy |

# Elementary Science Bingo

| Amphibian | Lungs | Clouds | Sound | Sun |
|-----------|-------|--------|-------|-----|
| Seasons | Skin | Equator | Plant | Winter |
| Blood | Volcano | | Rocks | Energy |
| Smell | Birds | Earth | Electricity | Year |
| Light | Water | Precipitation | Hair | Spring |

# Elementary Science Bingo

| Blood | Touch | Fish | Lungs | Day |
|-------|-------|------|-------|-----|
| Electricity | Light | Ocean | Winter | Equator |
| Earthquake | Mountain | | Water | Clouds |
| Heat | Moon | Year | Simple Machines | Skin |
| Nerves | Brain | Reptiles | Summer | Volcano |

# Elementary Science Bingo

| | | | | |
|---|---|---|---|---|
| Sound | Sight | Birds | Light | Seasons |
| Spring | Blood | Nerves | Rocks | Brain |
| Earthquake | Skin | | Fish | Temperature |
| Heat | Insects | Ocean | Year | Planets |
| Water | Lungs | Simple Machines | Touch | Autumn |

# Elementary Science Bingo

| Rocks | Autumn | Lungs | Clouds | Sight |
|-------|--------|-------|--------|-------|
| Electricity | Summer | Temperature | Reptiles | Winter |
| Touch | Day | | Air | Sound |
| Planets | Water | Nerves | Hair | Fish |
| Star | Sun | Moon | Mountain | Precipitation |

© Barbara M. Peller

# Elementary Science Bingo

| Sight | Touch | Summer | Lungs | Air |
|-------|-------|--------|-------|-----|
| Amphibian | Energy | Plant | Nerves | Seasons |
| Volcano | Equator | | Smell | Insects |
| Heat | Wind | Hearing | Hair | Water |
| Heart | Mountain | Sun | Year | Fish |

# Elementary Science Bingo

| Electricity | Autumn | Plant | Lungs | Magnet |
|---|---|---|---|---|
| Volcano | Fish | Birds | Clouds | Mammals |
| Skin | Moon | | Touch | Simple Machines |
| Nerves | Reptiles | Water | Heat | Mountain |
| Smell | Sun | Precipitation | Blood | Hair |

# Elementary Science Bingo

| Star | Planets | Fish | Taste | Blood |
|------|---------|------|-------|-------|
| Seasons | Summer | Sound | Clouds | Air |
| Birds | Winter | | Mammals | Equator |
| Water | Heat | Hair | Insects | Day |
| Sun | Brain | Touch | Skin | Plant |

# Elementary Science Bingo

| Amphibian | Touch | Reptiles | Taste | Precipitation |
|-----------|-------|----------|-------|---------------|
| Autumn | Sight | Moon | Electricity | Insects |
| Planets | Blood |  | Hearing | Mammals |
| Skin | Sun | Water | Brain | Hair |
| Magnet | Wind | Mountain | Nerves | Fish |

# Elementary Science Bingo

| | | | | |
|---|---|---|---|---|
| Amphibian | Sight | Day | Touch | Clouds |
| Rocks | Precipitation | Plant | Seasons | Mammals |
| Equator | Sound | | Blood | Skin |
| Magnet | Hearing | Water | Brain | Volcano |
| Heart | Smell | Sun | Summer | Wind |

© Barbara M. Peller

# Elementary Science Bingo

| Smell | Plant | Touch | Simple Machines | Fish |
|---|---|---|---|---|
| Insects | Magnet | Electricity | Birds | Air |
| Volcano | Clouds | | Hearing | Water |
| Sound | Heat | Wind | Sun | Winter |
| Precipitation | Day | Equator | Light | Heart |

# Elementary Science Bingo

| | | | | |
|---|---|---|---|---|
| Fish | Touch | Planets | Seasons | Sound |
| Nerves | Summer | Clouds | Sight | Amphibian |
| Magnet | Hearing | | Winter | Smell |
| Blood | Taste | Heat | Sun | Water |
| Equator | Light | Simple Machines | Wind | Heart |

# Elementary Science Bingo

| Planets | Birds | Touch | Sight | Energy |
|---------|-------|-------|-------|--------|
| Magnet | Hearing | Electricity | Water | Air |
| Ocean | Wind | | Sun | Smell |
| Sound | Autumn | Plant | Heart | Insects |
| Blood | Winter | Fish | Star | Equator |

# Elementary Science Bingo

| | | | | |
|---|---|---|---|---|
| Rocks | Sight | Sound | Touch | Birds |
| Energy | Fish | Hearing | Seasons | Winter |
| Wind | Skin | | Equator | Nerves |
| Year | Star | Moon | Sun | Water |
| Taste | Earthquake | Blood | Heart | Magnet |

# Elementary Science Bingo

| Fish | Sight | Sound | Electricity | Earthquake |
|------|-------|-------|-------------|------------|
| Magnet | Nerves | Plant | Equator | Star |
| Volcano | Hearing |  | Air | Touch |
| Energy | Heat | Temperature | Sun | Water |
| Amphibian | Clouds | Heart | Autumn | Wind |

© Barbara M. Peller

# Elementary Science Bingo

| Day | Touch | Seasons | Earthquake | Water |
|---|---|---|---|---|
| Insects | Sight | Planets | Winter | Air |
| Magnet | Blood |  | Equator | Plant |
| Heart | Autumn | Temperature | Sun | Hearing |
| Heat | Reptiles | Wind | Fish | Sound |

© Barbara M. Peller